Arctic Hare

by Grace Hansen

Abdo Kids Jumbo is an Imprint of Abdo Kids
abdobooks.com

abdobooks.com

Published by Abdo Kids, a division of ABDO, P.O. Box 398166, Minneapolis, Minnesota 55439.

Abdo Kids Jumbo™ is a trademark and logo of Abdo Kids.

Printed in China

102019

012020

Photo Credits: iStock, Minden Pictures, National Geographic Image Collection, Shutterstock, SuperStock

Production Contributors: Teddy Borth, Jennie Forsberg, Grace Hansen
Design Contributors: Dorothy Toth, Pakou Moua

Library of Congress Control Number: 2019941217

Publisher's Cataloging-in-Publication Data

Names: Hansen, Grace, author.

Title: Arctic hare / by Grace Hansen

Description: Minneapolis, Minnesota : Abdo Kids, 2020 | Series: Arctic animals | Includes online resources and index.

Identifiers: ISBN 9781532188893 (lib. bdg.) | ISBN 9781532189388 (ebook) | ISBN 9781098200367 (Read-to-Me ebook)

Subjects: LCSH: Arctic hare--Juvenile literature. | Arctic rabbit--Juvenile literature. | Rabbits--Juvenile literature. | Zoology--Arctic regions--Juvenile literature. | Arctic--Juvenile literature.

Classification: DDC 599.322--dc23

Table of Contents

The Arctic

The Arctic is the northernmost part of Earth. It is made up of land, the Arctic Ocean, and the **sea ice** that floats on it. The weather there is freezing cold. Any animal that lives in the Arctic is tough!

Arctic Hares

Arctic hares are made to survive in their very cold and snowy homes. They have large, long hind feet. These act like snowshoes and help them move easily over snow.

Their winter coats are snow-white in color. This helps them blend in with their surroundings. It keeps them safe.

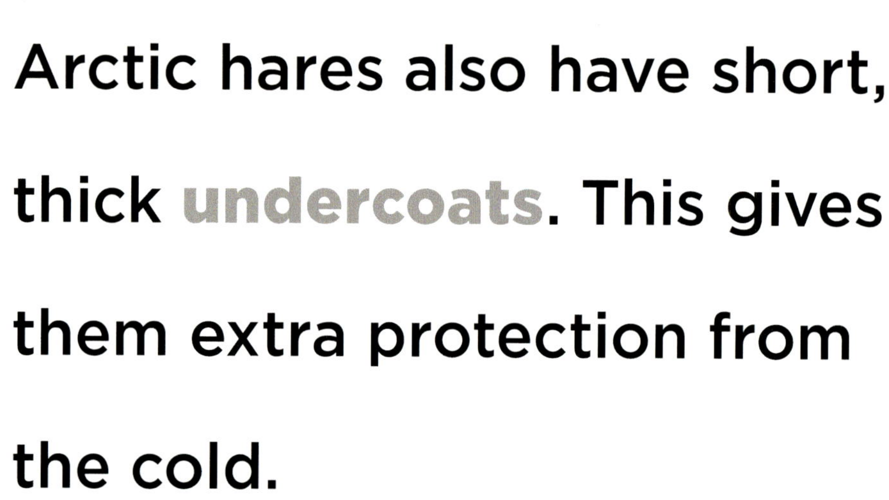

Arctic hares also have short, thick **undercoats**. This gives them extra protection from the cold.

Their ears are short. This keeps in body heat.

Arctic hares have sharp claws on their feet. They use these to dig **shelters**. They sometimes huddle with others to stay warm.

They also use their claws to dig for food. In wintertime, it can be hard to find food. They find twigs and other plants to eat under the snow.

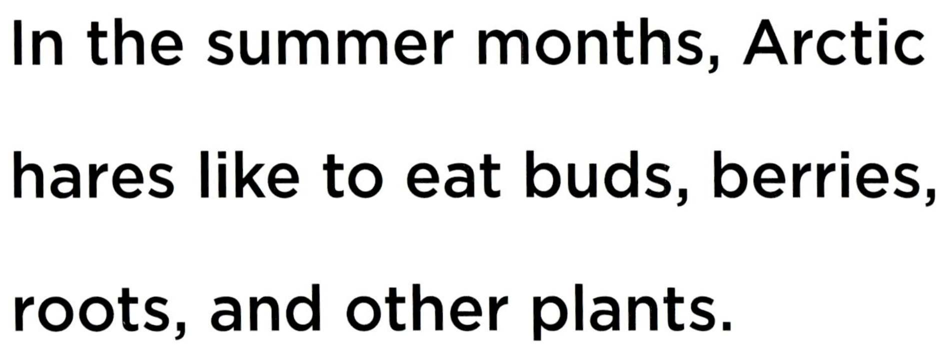

In the summer months, Arctic hares like to eat buds, berries, roots, and other plants.

Baby Arctic Hares

Females give birth one time a year in spring or summer. **Litters** have 2 to 8 **leverets**. Leverets grow quickly. At about 5 weeks old, they can leave the nest to find food.

More Facts

- Arctic hares are the largest species of hare.
- If an Arctic hare senses danger, it stands on its hind legs to look around. If scared, it can run up to 40 miles per hour (64 kmh) to escape.
- Arctic hares have to avoid animals such as ermine, Arctic foxes, and snowy owls.

Glossary

leveret – any hare that is under one year old.

litter – a group of young animals born to one mother at one time.

sea ice – frozen ocean water that is typically covered with snow.

shelter – a place or structure that gives protection against weather or danger.

species – a group of living things that look alike and can have offspring together.

undercoat – the short hairs growing close to an animal's skin, covered by longer hair or fur.

Index